May 2024

sorry if there are any errors. We don't have time, that's why I wrote it as best I can - there is no time for corrections.

Dedication:
For my Beloved Son who is the light of my life You are an infinite force for me to work in preparing for you a world where you will be complete y safe.

and

For an amazing and at the same time extremely humble man who, despite pain and rejection, constantly strives for his goal of saving the world. He sacrificed his own life to save the lives of all of us. Let this book be a testimony of your unwavering will, Wiesław Mróz. Let the words contained in these pages be a bridge between your extraordinary knowledge and our common hope. Thank you for your dedication. Thank you that you are.

Acknowledgments
Thank you, Mom - for your kindness, help, care and love.
Thank you for being here - you hold the fruit of my life in your hands.

>>> . . . ___ ___ ___ . . . <<<

Warning against copying the ebook:
Any unauthorized copying, distribution or use of the content of this e-book is strictly prohibited. Violators of these copyrights will be subject to severe fines in accordance with applicable law.

May, 2024, Katowice, Poland, planet Earth
ATTENTION! ALL INFORMATION CONTAINED HERE IS TRUE

ABOUT THE AUTHOR

The only thing I want to write about myself here is that there is nothing more important to me than saving the world. It's been six years since I devoted myself entirely to this extremely difficult and exhausting role, which is in fact a moral obligation of all people living in our common home - Earth. When Wiesław Mróz - engineer, discoverer, genius - called me for the first time over three years ago, I was not ready. I failed. He then asked me for help, he had one goal - to share his knowledge with humanity... After a few days, I lost my phone, and thus my contact with him, and I waited for 3 years for him to call. And so it happened - he called at the beginning of January this year and for 4 months, every day, I have been working for the Polish Science Foundation, engineer Wiesław Mróz - whose health condition is deteriorating, who, as he says himself, has little time - and the doctors only say one thing - death. I will add that the disease is the result of many years of total dedication to science, research and hard work that he did for all of us and for our planet Earth. This e-book was created with the hope of reaching all ends of the earth and all hearts, even the most hardened ones.

INTRODUCTION - URGENT CALL

When the first rays of the sun break the night darkness, the awakened Humanity illuminates the darkness of unconsciousness and apathy.

This is the moment when every beating heart on Earth must answer the call. A call to unite to defend our common only home - Earth.

I present to you my debut book in the form of an e-book, which is more than just words. It is a manifesto full of hope, a concrete action plan that has the potential to change the course of history.

Let this ebook be the spark that ignites the flame of change. Let it be a bestseller that will not only find its way into the hands of many, but also into the hearts of everyone.

As a woman, mother and activist, I have experienced loss and struggle. But I also saw what is possible when people join forces for a common cause.

As a citizen of the Earth, I want this book to be a bridge between what we know and what we can achieve. It is not just a collection of words - it is our historic chance for survival. Let it be a bridge between us - people - and the beginning of historic changes that we will make together.

With every word, with every page, we are getting closer to our goal. Let this e-book become a bullet that will pierce the wall of indifference and ignite a spark of change in everyone who reads it and hears about it!

CHAPTER 1
INTRODUCTION

Have you ever wondered how much money has influenced our lives? Are we truly free or enslaved? In a world where the **money fetish** seems to dominate over basic human values, we are faced with the question about the **remains of human morality** and more... How could we allow our actions to lead to the poisoning of Nature, which we are a part of and thanks to which we live? As Humanity, we have completely strayed from the path of understanding and compassion, and the consequences are terrifying. Money has become a tool of control that shapes our lives, society and the world. From an early age, we learn that success is measured by the size of our bank account, and our value as people is often assessed through the prism of wealth. In this reality, money has become not only a means of life but also an end in itself. Did he give us freedom? Or are we enslaved by the need to constantly acquire it? Working from dawn to dusk, endless debts, pressure and consumerism - all this creates the image of life in the shadow of money. It is concentrated in the hands of a few, while the majority of humanity struggles with poverty. The system that was supposed to serve as a tool for exchange has become a mechanism that deepens divisions. Money has become the driving force of an economy that does not take into account the consequences for the planet. The pursuit of wealth contributes to the ecological crisis.

In the face of war, hunger and poverty that constantly affect millions of people around the world, we must ask ourselves about our common responsibility. Can we continue to ignore the suffering of others when our actions – or lack thereof – contribute to their plight? Our responsibility does not end at our borders - it extends to all of humanity and future generations.

In today's world where money is ubiquitous, it is difficult to imagine life without it. However, more and more people are beginning to realize that the current financial system has taken away our freedom and contributed to social inequality. Money has become a chain that chains individuals to a constant struggle for survival instead of allowing them to realize their full potential. It is a source of inequality, an encouragement to excessive consumerism and a cause of environmental degradation. Money in its current form often leads to destructive actions. In the pursuit of profits, corporations and individuals ignore the damage they cause to the environment. This approach contributes to the overexploitation of natural resources, pollution of the environment, air and water and acceleration of climate change, which in turn affects our health and the well-being of all living things on Earth. Money has a huge impact on our planet, but we have a choice in how we use it. We can direct capital towards this project that serves us all. Each of us has a role to play in this change. Let us imagine a society in which cooperation replaces competition and the common good is more important than individual profit. In such a world, the exchange of goods and services could be based on mutual trust and needs, rather than monetary value. A world where the value lies in us, in natural resources, community and ecological balance. We can achieve this if we take the right actions. Eliminating money from the market may seem like a radical idea, but it is a viable concept that is gaining popularity as more people seek alternative paths to freedom. This is not just a dream of utopia, but a real vision of the future that requires courage, imagination and determination to become a reality. This is a goal that we can strive for through education, cooperation and action. This is the message that the Wiesław Mróz Scientific Foundation wants to convey to the world.

role of Education

In the face of global challenges, disasters and climate change, it is crucial to raise the level of education and ecological awareness around the world. Education is a powerful tool that can transform society and bring about lasting change. Global education should be available to everyone, regardless of age or origin. Education should help us understand ecological processes and their impact on our everyday lives. By investing in education, we invest in a better and fair future for all.

Saving the Earth is a task that requires global cooperation and commitment. The challenges are enormous, but with collective efforts we can achieve our goals. We must act quickly and decisively to ensure humanity's survival and to ensure a healthy planet for future generations.

Hunger and poverty are plagues that have accompanied humanity for centuries. Nowadays, despite enormous technological progress and wealth that would seem to be available to everyone, these problems are still present. How is it possible that in the 21st century there are still places where basic necessities of life are lacking?...

The **money fetish** has become a new idol before which both individuals and entire nations kneel. In the pursuit of profit, humanity has forgotten basic humanitarian values. Humanity has lost the meaning of life...

Industrial progress has brought us many benefits, but at the same time it has caused the degradation of our environment. Revealing Knowledge, if heard by Humanity, will lead to great changes that will occur naturally because understanding the natural world, learning the truth about the origins of our life and evolution will restore Consciousness to Humanity. When People find out who they really are and what everything around them is, Humanity will consolidate and the Rescue System, which I will mention later, will naturally be activated by Humanity automatically.

We are no longer defenseless in the face of all these challenges. They no longer scare us, they connect us and become the glue that will connect us all, because when the wave of help, goodness and knowledge starts, it will grab everyone and there is no other way.

Equality in access to basic resources such as clean water, healthy food and education are crucial to the development and well-being of every person. It is true that every person, like a flower, has the potential to grow and develop, but they need appropriate conditions for this.

Completely new opportunities are opening up to all of us, but in order for them to be realized, financial support is necessary. Why is it so important? The Foundation is committed to creating and distributing educational materials that help develop global competences in students around the world. Achieving these goals requires significant financial resources. Advertising activities are necessary for the foundation's message to reach the widest possible audience around the world. Global advertising costs are enormous, but necessary to increase awareness and social engagement. Financial support enables the foundation to conduct research, develop technologies and implement projects that have a far-reaching impact on our health and the environment. Without appropriate resources, even the best goals and initiatives may remain unrealized. Financing external sources will ensure the foundation's independence from commercial and political influences, which is necessary for the objectivity and credibility of research. Financial support allows for the exploration and development of new technological solutions that will lead to scientific breakthroughs. It is an investment in the future that will benefit the entire society. Every contribution, regardless of size, matters and is a step towards a better tomorrow. Therefore, I encourage you to support this fascinating foundation to enable it to continue what it was created to co - to save and change the world for the better. (details about financial support are at the end of the e-book).

Revealing Knowledge will revolutionize literally EVERYTHING - our approach to life, our relationships and our understanding of Nature and ourselves, if only you do not ignore it, because your task is to pass this information on, so that every citizen of the Earth learns about the existence of this Knowledge, which is our greatest treasure.
Now, thanks to Discovery Knowledge, all these lost values will be rebuilt. We all long for love, acceptance, understanding and peace…. This is what

each of us desires, and this book has been written for you all to hear what is coming - the REVOLUTION - through the global release of Revealing Knowledge, which will be provided on the website www.goliatvictory.com, which is under construction.

Through education and increasing awareness, we can create paradise on Earth.

EDUCATION is the foundation on which we can build lasting changes. Sustainable development is not just a slogan - it is a necessity for the survival of our planet

It is extremely important to combine knowledge and technology with social and environmental responsibility. This connection has the power to transform our world into a place where everyone has a chance for a better life, and the natural environment is protected, respected and flourishes with us...

Poverty and environmental degradation have led to the breakdown of communities. Now you know that we can build a fairer society where everyone has a chance to live with dignity.

Science and society are not isolated from each other - they can and should cooperate for the common good.

Education is not only about transmitting knowledge, but also about shaping attitudes and values.

Knowledge and education are key, but cooperation and responsibility are equally important.

The economy does not have to be based solely on profit, and let the world be ruled by love and truth!

Rescue operations under the supervision of the Foundation are not only necessary, but also possible to implement. Each of us can contribute to these activities - by supporting the foundation, changing our own habits and active participation. To prevent further damage, we must take immediate action. We must act together. Global challenges require global solutions. Earth Rescue Actions are a must. The foundation I will talk about later is ready to carry out these activities, but it needs support.

CHAPTER 2 ALARM - LAST CALL FOR EARTH

As you read this, our planet is on the brink of the abyss. The Earth, this blue jewel hanging in the cosmic void, screams for help, although its cry is often lost in the noise of everyday life and non-human ignorance.

For centuries, people have treated Her as an inexhaustible source of resources, regardless of the scars that our actions leave behind. Now, as climate change transforms our ecosystems beyond recognition and extreme weather events become the "new normal", we can no longer pretend that the problem does not exist.

Consequences of ignoring environmental problems

If we don't act, the consequences will be catastrophic. Sea level rise is threatening coastlines and islands, causing population migration and the loss of unique ecosystems. Extreme weather events such as hurricanes, floods and droughts destroy cities, crops and people's lives. The extinction of species, including those that are key to the balance of ecosystems, leads to the destabilization of entire networks of life

Doom in the Shadow of Progress

In an era when technology is developing faster than ever before and our daily lives are becoming more automated and digital, it is easy to forget about the price our planet is paying for it. The progress that has brought us so many comforts and conveniences also has its dark side - the destruction of the natural environment, which is happening almost unnoticed.

*Ecological Effects of Progress**

Our rivers and oceans are polluted with chemicals and plastic. The air we breathe is becoming more and more toxic. There are fewer and fewer forests, and with them animal and plant species are disappearing. Climate change leads to extreme weather events that threaten the lives of all people.

Consumerism and the Environment

We are a consumer society. We are encouraged to buy newer and newer products, often without considering their real need or impact on the environment. The fashion for "being eco" is often only superficial, and environmental activities come down to symbolic gestures that have more to do with marketing than real care for the planet.

Change is Necessary

We can no longer ignore the consequences of our progress. We must act now, before it is too late. This is a process that requires the individual commitment of each of us - without exception. Changes in our thinking, our habits, our approach to life and consumption. We must learn to live in harmony with nature, not to its detriment. We are the hosts of our Earth.

Here's an introduction to the most important issues and why immediate action is essential:

1. *Climate Change*: Our planet is experiencing rapid climate change, leading to rising sea levels, melting glaciers and extreme weather. This is not only a threat to natural ecosystems, but also to economies and human communities.

2. *Environmental pollution*: Air, water and soil pollution has a direct impact on human and animal health. Industrial waste, plastic and other toxins accumulate in our environment, causing long-term damage.
3. *Deforestation*: Deforestation on a massive scale leads to loss of biodiversity and is one of the most important factors contributing to climate change. Forests are our "lungs of the planet" and protecting them is crucial to the health of the Earth.
4. *Resource Depletion*: Overuse of natural resources such as water, minerals and fossil fuels leads to their depletion. We must promote sustainable practices to ensure their availability for future generations.
5. *Loss of biodiversity*: Species are becoming extinct at an alarming rate. Every species lost is an irreversible loss and affects the delicate balance of ecosystems.
6. *Overproduction and consumerism*: Our society produces more than it can use, which wastes resources and increases greenhouse gas emissions.
7. *Soil degradation*: Intensive farming, overgrazing and soil erosion lead to loss of soil fertility. Soil is the foundation of our food chain and we must protect it.
8. *Water resources*: Overutilization and pollution of water resources leads to water crisis. This affects human health, agriculture and aquatic ecosystems.
9. *Changes to ecosystems*: Climate change and human activities are disrupting the balance of ecosystems. This can lead to their collapse and loss of ecosystem services such as pollination and air purification.
10. *Threats from space*: humanity is not aware of the cosmic impact on our climate and is not aware that space is the main culprit of climate change on Earth. It would be worth investing in early warning systems.
11. *Demographic pressure*: The growing human population puts pressure on natural resources. We must strive for sustainable development to ensure the survival of our planet.
12. *Changes in biogeochemical cycles*: Human activities are changing biogeochemical cycles such as the carbon and nitrogen cycle. This has far-reaching effects on the environment.

Communities around the world must turn their attention to the growing and increasing climate problems and other ecological challenges. It takes education, innovation, collaboration.and dedication.

To address these challenges, we must act now. Immediately !Together! Global! Although time is against us - we still have it, we can still use it and get ahead of the catastrophe, master the current anomalies and introduce global education, which will result in the creation of a completely new world. A world that is completely free, healthy, clean and safe for all people, because I know how to do it and it is not a utopia. This is a real thing that, thanks to your involvement, may come true not in thousands of years, but now! In front of our eyes and the eyes of our beloved children.
Ignorance - the invisible enemy of nature

Ignorance in an ecological context is like the *invisible enemy of nature*. Ignorance can be seen as a lack of awareness or understanding of the impact that space and human activities have on the natural environment. Unconscious action or lack of action is nature's greatest enemy. People too often ignore environmental threats. The reasons vary, such as lack of direct impact on their daily lives, political or economic beliefs. People often mistakenly believe that they cannot change anything... Unfortunately, the consequences of this lack of awareness affect us all.

Climate change, loss of biodiversity, environmental pollution, humanitarian crises, ecological disasters, hunger and poverty are just a few examples of the real consequences of ignoring ecological problems.

The psychology of ignorance indicates that people may ignore environmental threats for a variety of reasons, such as a lack of immediate impact on their daily lives, political or economic beliefs, or even information overload.

How can we break this barrier of ignorance? Here are some strategies to increase environmental awareness:
1. *Education*: Introducing environmental education at all levels of education is crucial. We need to understand how our actions affect the environment and what the consequences are.

2. *Engaging local communities*: Local communities can work together to protect their environment. Community initiatives such as beach clean-ups and tree planting can help raise awareness.
3. *Promote a sustainable lifestyle*: Let's choose products with a lower impact on the environment, reduce or eliminate plastic consumption and save energy.
4. *Media and Communication*: The media can play a key role in informing the public about environmental issues. Media campaigns and educational articles can increase awareness.
5. *Action*: Each of us can act. Small steps such as segregating waste, giving up household chemicals, and limiting water consumption are important.

Strategies to increase environmental awareness may include environmental education at all levels of education, engaging local communities in conservation activities, promoting sustainable lifestyles, and using the media to inform and inspire pro-environmental actions.

The role of the media and education is crucial in shaping ecological awareness. Media can serve as a platform to disseminate information and promote good practice, while education can equip people with the knowledge and tools necessary to make informed environmental decisions.

A call to action is essential to break the barrier of ignorance and initiate change. Each of us can contribute to environmental protection through everyday choices, support for ecological initiatives and active participation in social life aimed at promoting sustainable development.

CHAPTER 3 Awakening Consciousness/Planetary Consciousness

The awakening of consciousness at the planetary level is a process that begins with individual understanding and expands into collective consciousness. This global thinking combined with local action has the potential to lead to lasting change on a global scale.

Understanding the problems is the first step. Education and awareness are crucial for people around the world to understand the environmental and social impacts of their actions. This understanding can lead to changes in attitudes and behaviors.

Thinking globally means recognizing that although we live in different communities and cultures, we share one planet and common challenges such as climate change, poverty and inequality. Global thinking inspires us to seek solutions that serve the common good.

Local action is how individual and community initiatives can contribute to global change. Local environmental, educational, and social support projects, although small, can have a far-reaching impact when replicated around the world.

Global sustainable change can be the result of synergy between understanding problems, thinking globally and acting locally. When enough people become aware and take action, it can create a domino effect leading to widespread change.

In this context, media and education play a key role in disseminating knowledge and inspiration for action. Through awareness campaigns, educational programs and sharing success stories, they can motivate and mobilize people to work for a better future for our planet[56].

This *call to action* is not only a moral imperative, but also a practical necessity to ensure a lasting and sustainable future for the Earth and its inhabitants. Each of us has the power to contribute to this change, regardless of where we are.

Awareness is the first step - analysis of the psychological and social aspects of climate change denial and ways to break the barrier of unconsciousness

Awareness of climate problems is indeed the first step to solving them. *An analysis of the psychological aspects of climate change denial* indicates that people may not accept the facts of climate change due to fear of the consequences it may bring to their lives and future[4]. Denial is often driven by an emotional response to the climate crisis, rather than a lack of information or ill will.

And the *social aspects of denial* include the impact of climate change on income inequality and migration. Poor countries, which have contributed little to climate change, are already more affected by the negative consequences of change than rich countries[4]. This may lead to a feeling of injustice and lack of motivation to act.

Ways to break down the barrier of unconsciousness may include education and communication tailored to the values and beliefs of different social groups. It is important to talk about climate change in a context that is familiar and understandable to the individual or community[5]. Adaptation and mitigation actions should be promoted as ways to improve the quality of life and ensure safety for future generations.

In summary, breaking the barrier of unconsciousness requires understanding the psychological and social causes of denial, as well as creating communication and education strategies that are effective and empathetic. This combination of awareness, education and action can lead to global lasting change.

Ecocentric awareness

CE is an attitude in which respect and care for the natural environment take a central place. Unlike anthropocentrism, which puts humans at the center, ecocentrism recognizes that all life forms and ecosystems have their own value and right to exist, regardless of their usefulness to humans.

In practice, a person with ecocentric awareness will strive to coexist harmoniously with nature, avoid overexploitation of natural resources, and promote activities that support sustainable development. This approach emphasizes the need to protect biodiversity and maintain ecological balance for the benefit of the entire planet, not just humanity.

Shaping and developing ecocentric awareness is a process that includes several key aspects:

1. **Education and Knowledge**:
    The basis is ecological education, which raises the awareness of individuals and communities about ecosystems, environmental threats and the effects of human activity[2]. Educational programs should be science-based and include lectures, workshops and educational trips.

2. **Behavior Change**:
    Environmental awareness requires changing everyday behavior. Promoting good solutions, such as the use of renewable resources, reducing plastic consumption, or saving water and energy, are practical steps towards ecocentrism[1].

3. **Social Involvement**:
    Local activities and social initiatives are of great importance for shaping ecological awareness. Participating in environmental protection projects, organizing clean-up campaigns or participating in ecological groups strengthens bonds with nature and motivates us to take further actions[2].

4. **Role of Teachers and Parents**:
    Teachers and parents play a key role in the environmental education process. Providing knowledge in an attractive way, building pro-ecological attitudes and encouraging environmental activities are necessary to shape awareness among children and young people[2].

5. **Practical Experience**:
    Direct contact with nature, such as trips to the forest, national parks or participation in gardening, allows you to better understand and appreciate the natural environment.

6. **Reflection and Self-Awareness**:

Understanding the impact our actions have on the environment is crucial. Reflecting on your habits and choosing more sustainable options is an important element of ecocentric awareness.

Let us remember that even the smallest step matters.

CHAPTER 4

BE A SPARK OF CHANGE and IGNITE OTHERS

Breakthrough Thoughts: Catalysts for Social Change

Throughout human history, it has not always been large armies or powerful corporations that have shaped our future. Often it was the power of one person, one idea that spread like a wave, changing society from the inside. In this chapter, we will look at some of these groundbreaking ideas and the people behind them to understand how an individual can become a catalyst for social change.

The Power of Ideas

An idea can be more powerful than armies. It is she who inspires, motivates and mobilizes people to act. When Martin Luther King delivered his famous "I Have a Dream", he not only expressed his personal desires, but also gave a voice to millions who wanted change. His words became a symbol of the fight for equality and justice, and his idea survived long after his death.

Brave Pioneers

Every era has its pioneers - people who are not afraid to set new paths. Maria Skłodowska-Curie, a two-time Nobel Prize winner, pushed the boundaries of science by discovering radium and polonium. Her courage and determination opened the door to new opportunities in medicine and physics.

Change by Action

It's not enough to just think about change; you have to act. Greta Thunberg, a young climate activist, showed that even the youngest have a voice and can influence global discussions. Her simple school strike has grown into an international movement that is forcing world leaders to pay attention to the climate crisis.

Conclusions

Now you see that each of us has the potential to be a catalyst for change. Regardless of age, origin or social status, we can influence the world around us. The stories of people who changed the course of history are proof that one person, one thought, can actually change the world.

The world we live in is like a canvas on which each of us can leave our mark. Many of us dream of a better tomorrow, but few people believe that they have the strength to make this dream come true. Every person has the power to change the world. You just have to dare - take the first step - join us. The world needs you.

Believe in Your Strength

The first step to making a change is believing in your abilities. Each of us has a unique set of talents and skills that can contribute to improving our environment. You have them too! Remember that even the biggest changes start with the first step. You don't have to change everything at once (but you can) - start with yourself, your community, your family, your friends, your workplace/school.

Don't wait for the perfect moment, because the perfect moment is now.

Within each of us lies the potential to be a force for change. Regardless of where we come from or what experiences we have, each of us can contribute to building a better tomorrow. Anyone can become a hero. I believe that the key to this change is empathy and understanding -

the ability to put yourself in the shoes of others and act with the common good in mind. Empathy can be practiced. You can arouse it in yourself.

Also remember that each of our gestures creates waves that spread far beyond our immediate surroundings.

CHAPTER 5
DON'T WAIT - ACT
Everyone's Action: Our Shared Responsibility to the Earth

In the face of global environmental challenges such as climate change, pollution and loss of biodiversity, the action of each of us becomes not only a moral imperative, but also a necessity for survival. In this chapter we look at why everyone's involvement is crucial to the future of our planet and how we can work together to protect it.

Global Challenges, Individual Actions

Although the problems facing Earth may seem huge and overwhelming, it is individual actions that have the power to bring real change. Each of us, by making conscious choices in everyday life, can contribute to reducing the negative impact on the environment.

Education for a Sustainable Tomorrow
Education is the foundation for understanding and action. By learning and teaching others about ecological problems and solutions, we build the foundations for a more conscious and sustainable future.

Small Steps, Big Changes

The power of small steps should not be underestimated. Saving water, separating waste, reducing energy consumption - all these are actions that, when undertaken by millions of people, have a huge impact on the condition of our planet.

Summary

Each of us has the power to influence the fate of the Earth. No matter where we are or who we are, our actions matter. Getting involved in saving the Earth is our common responsibility - for ourselves, for future generations and for all the wealth of life that shares with us this home called planet Earth.

Protection of life and nature as a moral duty of every person / moral imperative

Protecting the *environment* and *life on Earth* is indeed our moral obligation. There are many reasons why we should take care of our planet:
1. *Value of life*: Every form of life is valuable and deserves a chance to exist and develop.
2. *Ecological sustainability*: Biodiversity and healthy ecosystems are crucial to the sustainability of the environment we all use.
3. *Health and well-being*: A clean environment is the basis of our physical and mental health.
4. *Responsibility*: As inhabitants of the Earth, we have responsibility for its condition for future generations.

Education plays a key role in developing ethical attitudes towards the environment. It is through it that we can learn about the consequences of our actions and how we can contribute to protecting our planet. Education can also inspire active action and show practical ways to be more ecological in everyday life.

Protecting the planet is a shared heritage and responsibility. There are many activities in which we can contribute to this responsibility.

CHAPTER 6
APOCALYPSE

Deep in the depths of human consciousness lies the fear of the apocalypse - an end that is both terrifying and cathartic. In our times, the apocalypse does not take the form of fire from heaven or four horsemen. It is a silent but inexorable process that takes place in the background of our everyday lives.

The tearing of the apocalyptic veil is the moment when humanity must face the consequences of its ignorance. This is the moment when truths that have been ignored, downplayed or denied become impossible to get past. This is a time when scientific evidence on climate change, pollution and biodiversity loss can no longer be relegated to the margins of public debate.

We see how humanity has reached a point where the survival of our species and other forms of life on Earth is threatened. We explore how our everyday choices - from what we eat to how we travel to how we consume - contribute to the bigger picture of our planet's degradation.

But *ripping the apocalyptic veil* isn't just about revealing problems. It is also a call to action. This is a chance to awaken global awareness and understand that each of us has a role to play in saving our home - the Earth. It is a reminder that we have a moral responsibility to act not only for ourselves, but also for future generations.

Apocalypse: Breaking of the Apocalyptic Veil

This is a chapter that leaves no room for indifference. It is a manifesto for those who are ready to take up the challenge and become part of the solution. It is a warning, but also a guide on how we can work together for a better tomorrow.

This chapter is an invitation to deep reflection and active participation in creating a future that is sustainable, just and hopeful for all the inhabitants of our planet.

Apocalypse: Breaking of the Apocalyptic Veil

As humanity approaches the point of irreversible change on Earth, we increasingly feel the weight of our ignorance. The apocalypse is no longer just the subject of science fiction movies - it is a reality that we must face.

Breaking the Apocalyptic Veil - Awakening of Consciousness

Lifting the apocalyptic veil begins with awakening our consciousness. This is the moment when we stop ignoring warning signals and start taking action. Knowledge is the key - we need to understand what processes are taking place on our planet and what their consequences are.

Responsibility for the Future

Breaking the veil also means accepting responsibility for the future. Each of us has a role to play - from politicians and scientists to ordinary citizens. We must act together to minimize the impact of our actions on the climate, biodiversity and quality of life on Earth.

New Elections

Breaking the apocalyptic veil is a great opportunity for new choices. We need to redefine our goals and priorities. This is not the time to be selfish or short-sighted. We must think about future generations and what we leave as a legacy to them.

Summary

The breaking of the apocalyptic veil is the moment of truth. This is the time when we must decide what future we want for our planet. Will we choose wisdom and responsibility, or will we continue to ignore alarming signals? This is a question that each of us must find an answer to. It is a call to action that leaves no room for passivity. Our planet is our home - it's time to start taking care of it, and Poland plays a key role here!

Iskra from Poland - Poland on the World Stage

Poland, with its rich history and culture, has the potential to become a source of inspiration for other nations. The spark of change that will come from Poland has the potential to light the way to a better world. Poles have always defended their values and strived for freedom, and today we are really able to achieve it! Not through fighting, but through global education. Today, in a global society, Poland plays a key role on the international arena. Our history is full of heroes who fought for independence and human rights. We have a rich scientific and artistic tradition.

Poland, a country that has repeatedly become a symbol of courage and steadfastness, has all the predispositions to be a spark of change. Polish fighting spirit and determination can be an example to others on how to oppose injustice and strive for a better tomorrow.

But it is not everything. Currently, there is a man living in Poland - a genius of our times, who will revolutionize literally everything if only the world stops ignoring him. But before we get to this topic, I must add a few very important issues here.

CHAPTER 6
*** Light of Hope: Knowledge That Will Change Everything ***

In the darkness of the night, when the stars burned in the sky, hope was born. The Polish Science Foundation, hidden in the shadow of laboratories and equal only to itself, has discovered secrets that will shatter our previous beliefs. This discovery will not only change our knowledge, but also our hearts and souls.

This knowledge is like a ray of light piercing the darkness of ignorance. This is not a theory. This is the key to our future.

The Polish Science Foundation is the only one in the world that has groundbreaking knowledge and technologies that have the ability to heal our planet. This foundation plays a key role in what we call a "global rescue operation" because it has tools for this in the form of knowledge and ready-made projects. The Foundation's breakthrough discoveries cover a wide range of fields, from medicine and pharmacy, through neurobiology, chemistry to engineering and information technologies, as well as innovations in the field of energy and materials.

HOW DID THE WORLD REJECT HIM? Because he wasn't ready... But now he is, that's why you have this e-book in your hand that you can tell everyone you know about...., which should be read by every citizen of the Earth... Because this is where the beginning of all these beautiful, great changes takes place.... Maybe it is thanks to you that the world will accept and listen to this amazing man who has so much to tell us... He is our national treasure.... We cannot allow what he wants to give us to go with him.....

With great pride and admiration, I would like to introduce you to an extremely modest man, full of love, sensitivity and empathy, whom I loved with all my heart... and whose dedication exceeds the limits of reason and is deeply remembered.

engineer Wiesław Mróz (78) - president of the Polish Science Foundation, whose name I will not reveal (due to ongoing changes in the National Court Register), among others. discovered a source of energy that exceeds our wildest dreams. This is not another nuclear reactor or string theory. This is something that can change the course of history.
Appeal

I address this appeal to you, dear Reader. Let this knowledge not remain hidden. Let it become a light of hope for all people. Let our hearts beat in unison and our minds work together to understand this mystery. Now this is our common path.

CHAPTER 7

NEW ERA - TRUTH THAT LIBERATES

In the darkness of the laboratory, where science intertwines with dreams, something extraordinary happened. **A breakthrough discovery** that will shake the foundations of our understanding of the world. This is not another theory, it is not fiction. This is the **truth** that can liberate us and change the fate of not only humanity, but above all our planet Earth.

In the heart of Poland there is a Science Foundation, the president of which is 78-year-old engineer Wiesław Mróz - a friend close to my heart, with a heart bigger than everyone I know, whom I admire and respect. He is our greatest national treasure because he is the only one in the world who has knowledge that no one has ever heard of, and which, when it comes to light, will revolutionize literally everything. This is not some made-up fairy tale. This is a revolution that can transform our reality. The evidence is irrefutable, and the scientist, a genius who has been working in secret for many years, has one task: **to convey this knowledge to the world**.

A genius of his times, he has knowledge so groundbreaking that it exceeds the imagination of the modern world.

For years he has tried to pass on the knowledge he has accumulated over the years, because in his opinion it is the knowledge of humanity, so humanity must hear about it! And since you are here, know that Wiesiu - a man with unwavering faith in humanity, contacted every possible person, from scientists to world leaders, from small organizations to the largest institutions. But the answers were always the same: skepticism, disbelief, and sometimes even fear.

This knowledge is not a mere discovery; is the key to understanding life and the world in a way that can revolutionize all fields of science. This is our greatest legacy. But the people we met were not ready for this. This knowledge overwhelmed them and caused anxiety because it changed everything they took for granted. And I wasn't ready either when Wiesław met me for the first time.

Despite countless failures, Wiesław did not lose hope. I know that there are people out there who will understand the importance of his discovery and will be ready to act immediately.

This discovery is more than science - it is a mission that cannot be ignored.

And so, in the shadow of great discoveries and loud announcements, a new era was born in southern Poland. An era that could change everything (for the better, of course) if only the world were willing to listen...

And that's exactly why this ebook was written. To find you. You are ready to understand, to act and to change. This is a call to those who know that the future belongs to those who have the courage to dream and the strength to make these dreams come true.

We are part of a larger system: Our actions affect other people, the environment, the entire planet. Every choice, every decision matters.

Butterfly Effect: Small changes can lead to big results. What we do every day can have long-term effects.

Cooperation and solidarity: Together we can achieve more. Let's support each other and work together towards common goals.

An incredible opportunity for Poland:

As Poles, we have a unique opportunity to start a new era. We can build a spaceship that will open up space to us.

Our foundation has groundbreaking knowledge that can help heal our planet. Restoring it to its original form is a challenge, but we are ready for it.

Transfer of knowledge and involvement:

Our foundation wants to share this revealing knowledge with the whole world. This is not only our story, but the story of all people.

Each of you can be part of this story. Let's work together to create a better future for our children and grandchildren.

Dear compatriots, may our joint work bring positive changes. Let Poland become a symbol of innovation, cooperation and hope for the whole world.

CHAPTER 8
OUR NEW BEAUTIFUL AND HEALTHY WORLD

The vision of a new world is a call to action. Are you ready to contribute to creating this beautiful world together with other people? Sure. We are ready. We are worthy of it.

Welcome to a world where progress is no longer synonymous with destruction, where borders disappear and humanity creates the future together. It is a place where there is no government, police or army. Instead, there is a community in which everyone feels responsible for the common good. Let me tell you about this fascinating world.

1. **Safety and Harmony**:

   In the new world there is no need for a police force or army. People live in harmony, resolving conflicts peacefully. Aggression gives way to cooperation, and understanding other cultures and perspectives becomes the norm.

2. **No Social Inequalities**:

   The vision of a new world is also a fight against inequality. People share resources, ensuring that no one suffers from hunger or poverty. Education is available to everyone, and health does not depend on social status.

3. **Ecology and Cleanliness**:

   The Earth is a paradise where the air, water and soil are free from chemicals. People care about the natural environment by using renewable energy sources. All this allows for a long and healthy life.

4. **Knowledge and Awareness**:

   Developing awareness is the key to success. People are discovering how to avoid contracting viruses, and lying is becoming rare. Education is a priority and learning knows no boundaries.

5. **UFO Spaceship*

   A ship with an area of approximately 500 meters is not only a means of transport, but also a home. Safe, equipped with amazing technologies, it enables intergalactic travel. People discover new horizons by exploring the infinity of the universe.

(While the UFO project may be viewed as controversial, its potential to transform our approach to technology and nature is undeniable. It can inspire us to break boundaries and seek new, sustainable forms of energy that will have a positive impact on the future of our planet.)

6. **Cooperation and Empathy**:
   In the new world, people cooperate with each other instead of competing. Empathy is common, and everyone strives to understand the needs of others. They work together to solve global problems.
7. **Education and Creativity**:
   Schools are not limited to traditional subjects. Education also includes developing creativity, problem-solving skills and critical thinking. People learn throughout their lives.
8. **Health and Longevity**:
   Medicine has reached new levels. People live longer and diseases are rare. Medical technologies allow for quick diagnoses and effective treatment.

9. **Culture and Art**:
   The vision of a new world also involves the development of culture and art. People create beautiful works, are inspired by various traditions and express their emotions through music, painting, dance and literature.
10. **Discovering Space**:
    The UFO spaceship not only travels intergalactically, but also discovers the secrets of the universe. People explore other planets, encounter alien civilizations, and learn that we are part of something bigger.

11. **Technological Wonders**

   The new world is a place where technology serves humanity and the earth. People use advanced tools such as artificial intelligence, nanotechnology and biotechnology and tools created thanks to discovery knowledge, and so, for example, global problems such as climate change or hunger cease to exist. And things that we have never even dreamed of have their beginning of existence... because the universe, its exploration and science are INFINITE. And we are at the very beginning of this extraordinary journey....

12.. **Sharing and Reciprocity**:

   In this new reality, ideas of private property are evolving. People share resources, skills and knowledge, creating a system based on reciprocity and sharing. Exchange is no longer driven by the desire for profit, but by the desire to support each other and strengthen the community.

13. **Sustainable Development**:

   Ecology becomes the foundation of every decision. Technologies are developed and applied in a way that respects nature and promotes sustainable development. Humanity is working together to restore ecological balance and ensure a healthy planet for future generations.

14. **Education and Personal Development**:

   Education in Nowy Świat is available to everyone and focuses on personal development, creativity and abilities

It should be obvious to you by now that transformative discovery education is the key to a sustainable future.

CHAPTER 9
ENDLESS POSSIBILITIES THANKS TO PUBLISHING DISCOVERY KNOWLEDGE

In a new world where insightful knowledge is available to all, endless possibilities open up. It's fascinating that we have much more to discover than we thought.

Global Knowledge Exchange:

Dissemination of discovery knowledge is a process in which scientists, researchers and all of us share information. Thanks to this, everyone can benefit from the latest scientific achievements, regardless of origin or social status.

Technological Innovations:

The global publicization of scientific discoveries leads to a technological revolution. Inventions that were previously available only to a few are now becoming commonplace. This opens the door to new technologies that will change our world.

Disease Treatment and Longer Life:

Thanks to global access to groundbreaking medical knowledge, diseases are becoming rare. New therapies, the end of chemical drugs and diagnostic methods allow for a longer and healthier life. People live full of energy and vitality, do what they love and share it with others.

Education Without Borders:
   Schools and universities are becoming global education centers. Everyone has access to the best lecturers, materials and courses. Knowledge transcends national borders, and science becomes the common good of humanity.

Space Research and Planetary Exploration:
   Global publicity of space discoveries allows for joint space exploration. People explore other planets and discover the secrets of the universe. This is a time when boundaries do not exist.

   Dear people, together we can make this beautiful world come into being in our lifetime through our actions and sacrifice. Deep down you know it's true. Can you imagine anything more beautiful than uniting people around the world for such a noble purpose? I don't think so...

   For years, as a society, we have struggled with the consequences of our actions, which have irreversibly harmed our planet. But now, thanks to breakthrough discoveries, we stand on the threshold of a new era - an era in which science and technology go hand in hand with deep respect and care for Mother Earth.

Revealing Knowledge
   Engineer Wiesław Mróz, the founder of the foundation, after years of research and experiments, discovered methods that will change our approach to science and technology. His work, backed by solid evidence, shows that we can achieve sustainable development without harming the environment. We can heal our entire planet. This knowledge has the power to revolutionize our understanding of the world, evolution, life and the

cosmos, and we should do everything we can to contribute to it. It's an axiom.

This e-book is a manifesto that aims to unite people with similar beliefs and values who are ready to stand shoulder to shoulder to defend our planet. We are looking for pioneers, innovators, leaders - those who will help us convey this revealing knowledge to all humanity and start a rescue operation for our only home - the Earth.

Let's be united in action

At a time when the world is on the brink of irreversible changes, the need for unity becomes not only a desire, but a necessity. We live in an era where challenges such as climate change, poverty, social inequality and political tensions require a global response. These problems know no borders and affect every corner of the world, from the most remote villages to large metropolises

A historical task lies ahead of us. We have a chance not only to save the Earth, but also to initiate a change that will touch every aspect of our lives. The knowledge we have as a foundation can redefine our perception of progress, science, and even existence itself. I invite you to continue on your journey together. Let this ebook be a map that will lead us through unknown waters to the port where a better future awaits us. Together we can create a world where progress serves all, not just a few.

We are witnessing the progressive degradation of the Earth, which is our home and the host of all living beings. The time for change is NOW! Each of us has the power to contribute to saving our planet. Whether you are a scientist, engineer, artist or simply a citizen of this world, your actions matter. Global importance. Together we can create a new reality in which respect for nature and empathy will become the foundation.

When people unite in pursuit of a common goal, their actions become intuitive. The acquired knowledge and increased awareness allow for non-verbal communication - looks, gestures and even silence say more than words. In such harmony, every move is well thought out and every decision is shared. In a community where knowledge is shared, you don't need words to know what to do. Mutual understanding and shared values lead to synergy, where everyone brings something unique and the whole works like a well-oiled machine.

The beauty of working together and discovering new things lies at the heart of human progress. What we do together has the power to change the world. Let this chapter be a reminder that each of us has a spark within us that can ignite the flame of change.

"A system for saving Nature and Humanity from the progressive destruction"

This is a project that aims to unite people ready to build a better tomorrow. It is a system based on revealing knowledge, truth and the most important human values. At the center of this system we put children - our future, who we must prepare for life in the world we will hand over to them.

Using the "System for Saving Nature and Humanity from the Progressive Destruction" and good organization - Humanity has a chance to master the current cataclysms. The system provides specific solutions that will help stop or even completely eliminate climate anomalies. If the Rescue System is activated immediately, we still have a chance to survive.

Life, however, is not just about survival. It is also about development, learning and cooperation. The Polish engineer, explorer, inventor and scientist, whose works constitute the foundation of this e-book, believes that each of us has the potential to create and discover. It is thanks to this revealing knowledge, supported by evidence, that we can face the challenges of our times. The Polish Science Foundation is the guardian of this knowledge. She is the only one in the world who holds the key to understanding and repairing the world, which she wants to share with humanity. Together we can make choices that will have a positive impact on our planet and all its inhabitants.

This is our common mission - to create a new world in which every person, every child, every living being will have a chance for a dignified life.

But first this world must be saved... and healed... and this requires the individual commitment of each of us, our work and effort. Nobody will do it for us.

LIBERATION OF HUMANITY FROM THE YOKE OF SLAVERY

The knowledge presented will become the key to liberating Humanity from the current yoke of slavery and will revolutionize the life of Humanity for all time. It will show humanity the possibility of Life, which people only dream of. It is necessary to emphasize that the knowledge transferred to humanity will definitely help regain full awareness and facilitate the introduction of the expected changes. Legendary activist and author Joanna Macy wrote:

"If the world is to be healed through human efforts, I am convinced that it will be ordinary people who will do it. People whose love for life is greater than their fear."

Freedom, Health, Abundance and Security for Everyone

These four values form the foundation of our lives and well-being. Here's why they're so important:

1. **Freedom**:

Freedom is the right to choose, self-determine and act freely. Without it we cannot develop, create or fulfill our dreams. Freedom is the foundation of democracy and respect for human rights.

2. **Health**:

Health is the most precious treasure. Without it, we cannot enjoy life, achieve our goals, or care for others. Therefore, it is worth taking care of your health, both physical and mental.

3. **Abundance**:

Abundance means abundance, wealth and the ability to share with others. It is not only material goods, but also knowledge, love, friendship and respect. Striving for abundance allows us to enjoy life.

4. **Security**:

Security is a feeling of stability, protection and peace. Without it, we cannot plan the future or pursue our passions. Therefore, it is worth taking care of your own and others' safety.

Together we can achieve these values to create a better world for all.

Many of us wonder where the original freedom we felt as children has gone. Children have natural curiosity, creativity and the ability to act spontaneously. Their minds are not yet limited by social patterns and conventions. What seems impossible to adults is often everyday life for children. These are the remains of true freedom. But we can regain this freedom - fully. What took away people's freedom?
Money....

Money has become a tool of exchange, but also a source of inequality, corruption and exploitation. He limited us instead of serving us. As we grow up, we learn that money is necessary for survival, or is that really true? Does it have to be this way?

Consumer society places emphasis on possessions, appearance, and social status. People work harder and harder to get more money, but they often lose themselves in the process. Money has become a chain that has limited our choices and, in fact, our entire lives... Inequalities in access to financial resources, education, health care and other services have led to enslavement at the highest level. Low-income people are limited in their options and life choices. The modern financial system is based on debt and credit. People in debt are forced to work in unfavorable conditions or to perform jobs that do not suit their passions or skills. A consumer society promotes constant buying and possession. This has also led to enslavement when people are addicted to purchases and material possessions. High turnover in financial markets, speculation and manipulation have led to economic instability and loss of savings by ordinary people.

The modern job market often requires long working hours, which has led to an imbalance between work and personal life and has affected people's mental and physical health.

EARTH MAN CODE

This code is a proposal that can be an inspiration to reflect on what values and principles should guide people in life. These values are consistent with the universal principles of human rights, which are universal, inherent, inalienable, natural, inviolable and indivisible. It is a code that could serve as an ethical guide for humanity, a proposal of principles and values that can contribute to building a world for all of us:

1. *Unity*: We recognize that all people are part of a global community, united by common origins and common goals.
2. *Equality*: We strive for equality and justice for all, regardless of race, gender, age, religion or social status.
3. *Respect*: We show respect for all forms of life, understanding that each being has its value and place in the ecosystem.
4. *Harmony with nature*: We commit to living in harmony with nature, protecting its resources and promoting sustainable development.
5. *Peace*: We work for peace and conflict resolution through dialogue, cooperation and mutual understanding.
6. *Compassion*: We develop compassion and empathy by helping those in need and supporting each other in difficulties.
7. *Responsibility*: We take responsibility for our actions and their impact on others and the planet.
8. *Knowledge*: We strive to acquire knowledge and wisdom by learning from each other and from the world around us.

9. *Freedom*: We defend personal freedom and self-expression as long as it does not violate the rights of others.
10. *Legacy*: We pass on to future generations a world that is richer, not poorer, both culturally and naturally.
11. *Respect* - We treat everyone with respect, regardless of their background, beliefs or status.
12. *Empathy* - We strive to understand and share the feelings of others, living in harmony and compassion.
13. *Justice* - We act justly, striving for equality and fairness in society.
14. *Cooperation* - We cooperate with others for the common good and progress of humanity.
15. *Love* - We develop love and kindness towards ourselves and others as the foundation of healthy relationships and society.
16. *Harmony* - we strive to live in harmony with ourselves, other people, nature and the universe.

This code is a reminder of what really unites us and what values we carry deep in our hearts.

CHAPTER 10
THIS IS THE LAST CALL FOR EARTH

Last call for the Earth - a slogan that emphasizes the urgent need to act in the face of global challenges, such as climate change, environmental degradation, or threats from space. Here are some compelling arguments why now is the decisive moment to act:

*Climate anomalies: We are observing more and more extreme weather phenomena, such as heat waves, hurricanes and droughts. These anomalies are deviations from the long-term average and can lead to natural disasters, affecting the lives of millions of people.

*Progressing extinction: We have already lost many species of fauna and flora, and many more are endangered. Loss of biodiversity can destabilize the ecosystems on which our agriculture, forestry and fisheries rely.

*Threat from space: currently our planet, the threat of impact by asteroids and even large asteroids is very very high, and its effects could be catastrophic for the entire planet. It is therefore worth investing in early warning systems and defense plans.

*Astronomical discoveries: New discoveries, such as potentially habitable exoplanets, remind us of our responsibility to Earth as the only planet we know of that can support life.

*Lack of adequate safeguards: Despite technological progress, we still do not have sufficient safeguards against many global threats, including the effects of climate change.

It is now clear to you that action is necessary to prevent irreversible changes and secure the future of our planet for future generations. Each of us can contribute to change through commitment and action on behalf of our foundation, through education, changing consumer habits, supporting sustainable development and pushing policy makers to take bold and decisive steps to protect our common home.

Finally, I would like to emphasize that WE POLES have an extremely great chance and opportunity to change this sick world, we have a chance for reconciliation as a nation because this content was first written in Polish, so you Poles are the first to read it.

The Foundation was created for you, for all of us, so that we can make our dreams come true and restore paradise on Earth.
It has a *unique role in discovering new solutions*. Revealing knowledge moves. boundaries of knowledge and opens new cognitive perspectives. An example would be our patent for a virus with which we would never have to infect each other again. He is therefore a true *pioneer of changes* and

visionary of the future who not only educates but also inspires to search for new, previously unknown areas of science.

The limits remain only in your imagination, although after reading this e-book I believe that your imagination will no longer have any limits.

Each of us is an important link in the chain of change.
And so here it is - not the end, but the beginning.

This is the beginning of the revolution of which you are a part.
This is the beginning of great changes and beautiful moments.
We all write this story -
THE LIFE HISTORY OF HUMANITY

Accessories:
No. 1. Manifesto of Positive Speech
In every word I say
There is a force that shapes the world.
I choose words that reflect *respect*
love, *gratitude* and *peace*,
Because I know that what I send comes back to me.
Respect is the foundation,
Where I build bridges between hearts.
In every "good morning" and "thank you",
There is a power that connects and heals.
Love is a language that transcends borders,
It finds its place in words full of warmth.
When I say "I love", "I care" and "I support",
I create a space where everyone can grow.
Gratitude is the key that opens hearts.
"I am grateful" is a phrase that brings abundance.
By saying it, I attract prosperity,
And I'm learning to see beauty in everyday life.
Peace is the goal I strive for.

In the words "harmony", "peace" and "solution",
There is a force that soothes and heals,
Leading us towards a better tomorrow.
Let every word I say
It will be a reflection of these values.
Let them be the light that guides me,
And an inspiration to those who hear them...

no. 2 *Global Meditation for All Humanity*

I invite you to a joint meditation that will connect us all in one peaceful and positive act. This simple exercise can help create harmony and peace in the world. Here's how we can do it:

1. *Preparation:*
   Find a quiet place where you can sit or lie down comfortably. Close your eyes and relax.

2. *Breathing:*
   Take a few deep breaths. Imagine inhaling calm and exhaling any tension.

3. *Visualization:*
   Imagine that you are connected to every person in the world. We all breathe together, creating one big grid of energy.

4. *Intention:*
   Send your positive thoughts and intentions to all of humanity. You can think about peace, love, health and harmony.

5. *Mantra:*
   Repeat the word "peace" or "love" in your mind. Imagine these words spreading around the world.

6. *Thank you:*
   Thank you for this moment of peace and for the opportunity to influence the world with your thoughts.

7. *Ending:*
   Slowly open your eyes. Know that your meditation affects all of humanity.

Thank you for being part of this global meditation. Together we can create a more beautiful and peaceful world!

Supplement No. 3 *Deep Unification Meditation for Mother Earth

Sit comfortably or lie down, allowing your body to relax completely. Take three deep, calm breaths, inhaling through your nose and exhaling through your mouth.

1. Thanking the Earth

Focus your attention on your heart. Imagine a golden flower growing from your heart, a symbol of gratitude to Mother Earth.

With each breath, the flower blooms and its petals touch the earth, conveying your thanks to it for its constant gifts.

2. Connection with Nature

Imagine becoming a tree, with strong roots reaching deep into the earth's core.

Feel your roots absorb the nourishing juices from the soil as you become stronger and more united with every living thing.

3. Healing Visualization

Imagine a stream of clean, green energy flowing from your heart and spreading throughout the planet.

This energy reaches every corner of the world, healing polluted rivers, revitalizing damaged forests and protecting endangered species.

4. Call to Action

Think about specific actions you can take to help our planet.

It could be something simple like planting a tree, or something bigger like getting involved in a local environmental initiative.

5. Return to Consciousness
When you feel your meditation coming to an end, slowly begin to deepen your breathing.
Feel your body and mind energized and ready to take action for Mother Earth.

When you are ready, gently open your eyes, keeping love and concern for our planet in your heart.

no. 4
Compatriots scattered all over the world,
In every corner of the globe, where the day awakens and the night falls,
Where the heart beats with pride and the memory of ancestors has not faded yet,
I am turning to you - Poles, brothers and sisters in spirit and blood.

Let this call be a bridge that connects us,
Crossing the seas, mountains and borders that divide our paths.
It's time for us to unite our strength and hearts,
Let us face the challenges that lie ahead of us together.

Let us unite in defense of our mother - Earth,
Which has been feeding and sheltering us for centuries.
Let our actions, small and great, be our contribution,
In the protection of nature, in saving what is priceless and what is common.

Let each of us, wherever we are, become guardians of nature,
Let every gesture, every choice, reflect our care and love.

Let our Polish solidarity resound on all continents,
As an example that a nation of small numbers can make big changes.

Let this call be the beginning of a new era,
Where a Pole, regardless of place, becomes an ambassador of greenery.
Let our unity be an inspiration for other nations,
To create a world together where living in harmony with nature is the norm.

Compatriots, the time for action is now, not tomorrow, not someday,
This is our common mission, our duty, but also our great gift.
Let our Polish hearts beat to the rhythm of a green future,
And let our love for our homeland translate into love for the entire planet.

United in spirit, strength and determination,
Let us take up this challenge so that we can tell future generations:
"We acted together, as Poles, for the good of the Earth and life on it."

This call is a symbol of our unity and determination and that every Pole, wherever he is, feels part of this great mission.

Summary

In our journey through the pages of this e-book, we have covered a wide spectrum of topics - from the dark aspects of human morality to a ray of hope in the form of revealing knowledge that has the potential to revolutionize our world.

The remains of human morality in the face of the **money fetish** show us the image of a society that often puts material wealth above the well-being of others. However, this is not a hopeless picture. **Discovering knowledge** and technology open the door to a future in which we can live in harmony with nature and each other.

Hunger and **poverty** are problems that we can fight through education, international cooperation and responsible action. **Poisoning nature** requires us to react immediately and change the way we live and run businesses.

In this e-book, we have emphasized that each of us has a role to play. Whether we are scientists, politicians, entrepreneurs or citizens, our everyday choices and actions have an impact on shaping the future.

The invitation to action is clear - let's not wait for changes, let's be their initiators.

Let this e-book be a guide and inspiration for you to become part of the movement for a better tomorrow. Let's remember that the future starts NOW.

Final information

contact - Magdalena Machowska iskrazpolski44@gmail.com

facebook - https://www.facebook.com/magdallenamachowska

contact the foundation - pl.biuro@goliatvictory.com

website (under construction) - www.goliatvictory.com

The foundation is currently called FTL SHIP EARTLY - ultimately the Goliat Victory Foundation (changes in the National Court Register are ongoing, including the name of the foundation)

KRS No. 0000432782

Account number:

PL 30 1140 2004 0000 3102 8451 4451

titled "Donation"

Individual, corporate and private banking clients: BREXPLPWMBK
***

www.ingramcontent.com/pod-product-compliance
Lightning Source LLC
Chambersburg PA
CBHW082222220526
45470CB00010B/3270